1 MONTH OF FREE READING

at

www.ForgottenBooks.com

By purchasing this book you are eligible for one month membership to ForgottenBooks.com, giving you unlimited access to our entire collection of over 1,000,000 titles via our web site and mobile apps.

To claim your free month visit:

www.forgottenbooks.com/free1288961

* Offer is valid for 45 days from date of purchase. Terms and conditions apply.

ISBN 978-0-364-99160-2
PIBN 11288961

This book is a reproduction of an important historical work. Forgotten Books uses state-of-the-art technology to digitally reconstruct the work, preserving the original format whilst repairing imperfections present in the aged copy. In rare cases, an imperfection in the original, such as a blemish or missing page, may be replicated in our edition. We do, however, repair the vast majority of imperfections successfully; any imperfections that remain are intentionally left to preserve the state of such historical works.

Forgotten Books is a registered trademark of FB &c Ltd.
Copyright © 2018 FB &c Ltd.
FB &c Ltd, Dalton House, 60 Windsor Avenue, London, SW19 2RR.
Company number 08720141. Registered in England and Wales.

For support please visit www.forgottenbooks.com

Historic, archived document

Do not assume content reflects current scientific knowledge, policies, or practices

2
Ag82
2

(AIC-144)

UNITED STATES DEPARTMENT OF AGRICULTURE
Agricultural Research Administration
Bureau of Agricultural and Industrial Chemistry
Southern Regional Research Laboratory

SELECTED PUBLICATIONS

of the

Naval Stores Research Division

on

PRODUCTION, PROPERTIES, and USES of NAVAL STORES
(Pine gum, turpentine, rosin, and derivatives)

Naval Stores Station
Olustee, Florida

UNITED STATES DEPARTMENT OF AGRICULTURE
Agricultural Research Administration
Bureau of Agricultural and Industrial Chemistry

SELECTED PUBLICATIONS

of the

Naval Stores Research Division

on

PRODUCTION, PROPERTIES, and USES of NAVAL STORES
(Turpentine, rosin, pine gum, etc.)

The publications selected for inclusion in this list relate to production, uses, and composition of naval stores, including specifications, analytical methods, chemical and physical properties, and scientific and technical information concerning turpentine and rosin and some of their components and derivatives. Statistical reports [1] and patents are also included. The author of each of the publications listed was associated with the Naval Stores Research Division when the publication was issued. Only those publications believed to be of current interest are included.

In general, the publications marked "x" may be obtained from the Naval Stores Research Division, Naval Stores Station, Olustee, Florida. The supply of some of these publications, however, is insufficient to permit general distribution. Publications not marked "x" can generally be consulted in large public libraries. Photostat copies of any of these publications may be obtained from the U. S. Department of Agriculture Library, Washington, D. C., at a nominal cost. Copies of the United States patents can be obtained from the Commissioner of Patents, Washington 25, D. C., at a cost of 25 cents each (stamps are not accepted).

For the compilation and careful checking of this list of publications, acknowledgment is made to Hazel H. Fort, Marie M. Horner, and Dorothy B. Skau.

[1] See note under STATISTICS.

x Rosin strainers (drawing). J. O. Reed. NS-56. (1932).	152
x Turpentine dehydrator (drawing). J. C. Reed. NS-52. (1933).	157
Component distribution trend in commercial turpentine still operation. S. Palkin. Indus. and Engin. Chem. $\underline{25}$: 95-7 (1933).	158
x Turpentine separator (drawing). J. O. Reed. NS-72. (1933).	163
Naval Stores Station cleans dip barrels with steam. G. P. Shingler. Naval Stores Rev. $\underline{43}$ (27): 4 (1933).	167
x Preparation of special rosins. S. Palkin and C. K. Clark. Indus. and Engin. Chem. $\underline{26}$: 720-2 (1934).	182
x Retarding changes in turpentine during storage. W. C. Smith and H. P. Holman. Indus. and Engin. Chem. $\underline{26}$: 716-8 (1934).	183
x Rosin trough (drawing). E. L. Patton. NS-125. (1935).	198
x Summary of the 1934 distillation work at the Naval Stores Station, Olustee, Fla. G. P. Shingler and F. P. Veitch. Mimeographed. 4 pp. (1935).	201
x The outturn from several grades of gum containing small, medium and large quantities of chips and trash. G. P. Shingler. Mimeographed. 3 pp. (1936).	213
Care of government type still settings. K. S. Trowbridge and G. P. Shingler. Naval Stores Rev. $\underline{45}$ (49): 4, 15 (1936).	216
Production of clean gum rosin. W. C. Smith. Indus. and Engin. Chem. $\underline{28}$: 408-13 (1936).	220

PRODUCTION (continued)

Reference Number

x Rosin barrels, loss of moisture from filled barrels of rosin. G. P. Shingler and F. P. Veitch. Mimeographed. 3 pp. (1936). — 221

x Cooper's winch (drawing). NS-144. (1936). — 227

x Dehydration of turpentine. Sci. Amer. 154: 353 (1936). — 229

x Turpentine farmers lose by not removing chips from gum. Press Release. Mimeographed. 2 pp. (1937). Naval Stores Rev. 47 (2): 14 (1937). — 249

x Rosin lost in chips, rock dross and batting dross. G. P. Shingler, C. K. Clark, and N. C. McConnell. U. S. Bur. of Chem. and Soils. Mimeographed. MC-1. 4 pp. (1937). — 250

x Cross-sectional view of turpentine fire still layout at the Naval Stores Station. U. S. Bur. of Chem. and Soils. Mimeographed. C-12. 4 pp. (1937). — 257

x Government-type turpentine fire still setting, drawing and bill of materials. U. S. Bur. of Chem. and Soils. Mimeographed. C-13. 4 pp. (1937). — 260

x Directions for decolorizing turpentine by distillation. G. P. Shingler, E. L. Patton, and N. C. McConnell. U. S. Bur. of Chem. and Soils. Mimeographed. MC-17. 2 pp. (1937). — 261

x Rerunning rosin on a fire still. E. L. Patton and N. C. McConnell. U. S. Bur. of Chem. and Soils. Mimeographed. MC-18. 1 p. (1937). — 262

Gum grades and standards should be established. G. P. Shingler. Naval Stores Rev. 48 (18): 14 (1938). — 285

One and one-half pieces of cotton batting effect a saving to producers. G. P. Shingler. Naval Stores Rev. 48 (20): 21 (1938). — 286

Packing is important in naval stores products. J. O. Boynton. AT-FA Jour. 1 (9): 14 (1939). — 310

x Rosin straining. G. P. Shingler and E. L. Patton. U. S. Bur. of Agr. Chem. and Engin. Processed. ACE-1. 3 pp. (1939). Also under title: "The Vital Importance of Rosin Straining." Naval Stores Rev. 49 (20): 10 (1939). — 312

x Grades and yields of rosin and turpentine from crude gum. A. R. Shirley. U. S. Bur. of Agr. Chem. and Engin. Mimeographed. ACE-5. 3 pp. (1939). Naval Stores Rev. 49 (15): 4 (1939). — 314

PRODUCTION (continued)

		Reference Number
x	An experimental turpentine still. E. L. Patton. U. S. Bur. of Agr. Chem. and Engin. Processed. ACE-2. 4 pp. (1939). Naval Stores Rev. <u>49</u> (28): 4, 13 (1939).	315
x	Government-style fire still operated twenty-four hours a day. Naval Stores Rev. <u>49</u> (29): 22 (1939).	317
x	Turpentine still buildings and equipment. U. S. Bur. of Agr. Chem. and Engin. U. S. Dept. of Agr. Misc. Pub. 387. 44 pp. (1940).	325
x	Value of covering turpentine cups. R. V. Lawrence. Naval Stores Rev. <u>49</u> (49): 4 (1940).	329
x	How to discharge a turpentine still (Revised). U. S. Bur. of Agr. Chem. and Engin. Mimeographed. ACE-52. 2 pp. (1940).	336
x	How to charge a turpentine still (Revised). U. S. Bur. of Agr. Chem. and Engin. Mimeographed. ACE-53. 2 pp. (1940).	337
x	Directions for running crude gum on a turpentine fire still. U. S. Bur. of Agr. Chem. and Engin. Mimeographed. ACE-54. 5 pp. (1940).	338
	A study of the cost of production of naval stores. H. D. High. U. S. Bur. of Agr. Chem. and Engin. Mimeographed. ACE-59. 6 pp. (1940).	340
	Suggests gum buying on new weight basis. J. O. Boynton. Naval Stores Rev. <u>50</u> (31): 8 (1940).	346
	Avoid cracked still crowns. J. O. Boynton. AT-FA Jour. <u>3</u> (3); 10 (1940). Naval Stores Rev. <u>50</u> (35): 15 (1940).	348
	Protect your still plant from fire. J. O. Boynton. AT-FA Jour. <u>3</u> (3): 10 (1940). Naval Stores Rev. <u>50</u> (36): 13 (1940).	349
	How to stop leaks from the side sampler in a metal rosin drum. E. L. Patton. U. S. Bur. of Agr. Chem. and Engin. Mimeographed. ACE-72. 3 pp. (1941).	355
	Comments on the buying and selling of gum. G. P. Shingler and H. D. High. Naval Stores Rev. <u>50</u> (47): 8, 13 (1941).	361
x	Rosin cleanliness. G. P. Shingler and E. L. Patton. Oil and Soap <u>18</u>: 133 (1941). Also under title: "The Importance of Rosin Cleanliness and the Absence of Foreign Matter." Naval	

PRODUCTION (continued)

	Reference Number
Stores Rev. 50 (52): 8 (1941). Also under title: "Careful Handling Produces Cleaner Gum Rosin". (Rewrite). Paint, Oil and Chem. Rev. 103 (8): 16 (1941).	364
x Comparative summary of results with crude and cleaned gum in distillation work at the Naval Stores Station, Olustee, Fla. G. P. Shingler. U. S. Bur. of Agr. Chem. and Engin. Mimeographed. ACE-88. 2 pp. (1941). Naval Stores Rev. 51 (5): 16 (1941).	365
x Studies in the storage of turpentine gum for distillation. G. P. Shingler. Naval Stores Rev. 51 (6): 13 (1941). U. S. Bur. of Agr. Chem. and Engin. Mimeographed. ACE-89. (1941).	368
x A comparison of the effects of slash and longleaf gums on cup materials. R. V. Lawrence. U. S. Bur. of Agr. Chem. and Engin. Mimeographed. ACE-101. 2 pp. (1941). Naval Stores Rev. 51 (21): 13 (1941).	372
x Cleaning and painting turpentine cups -- revised. G. P. Shingler, R. V. Lawrence, and N. C. McConnell. U. S. Bur. of Agr. Chem. and Engin. Mimeographed. ACE-110. 3 pp. (1941).	373
x Upgrading oleoresin by a new process. J. O. Reed. Chem. and Metall. Engin. 48 (12): 68-70 (1941).	379
The adoption of the cup system by the naval stores industry. G. P. Shingler. Naval Stores Rev. 51 (36): 16, 20 (1941).	381
x Production of naval stores. U. S. Bur. of Agr. Chem. and Engin. U. S. Dept. of Agr. Misc. Pub. 476. 10 pp. (1942).	386
x Salvaging turpentine cups and accessories in the light of national defense requirements for metals. G. P. Shingler. U. S. Bur. of Agr. Chem. and Engin. ACE-136. (1942).	387
Turpentine distillation equipment and methods and the development of central stills. G. P. Shingler. Naval Stores Rev. 51 (40): 10, 16, 20 (1942).	388
Cup cleaning and painting method is demonstrated. Naval Stores Rev. 51 (43): 14 (1942).	389
Quart oil cans serve as turpentine cups and gutters. Naval Stores Rev. 51 (45): 12 (1942).	391

PRODUCTION (continued)

Reference Number

x Removal of metallic contaminants from pine oleoresin, washing with mineral acid. R. V. Lawrence. Indus. and Engin. Chem. <u>34</u>: 984-7 (1942).   408

Operation of the Olustee, Fla., Naval Stores Station process of gum refining. E. L. Patton and R. A. Feagan, Jr. Naval Stores Rev. <u>52</u> (41): 12 (1943). AT-FA Jour. <u>5</u> (4): 7 (1943). 418

x Preliminary results of tests on rosin and turpentine from gum produced by trees treated with chemical stimulants. Naval Stores Rev. <u>53</u> (47): 8 (1944). Also under title: "Spirits and Rosin from Treated Trees Tested in Olustee (Florida)." Jour. Amer. Farmers Assoc. <u>6</u> (5): 6 (1944).   436

x Further information on the effect of chemical stimulation on grades and cloudiness of rosin. Naval Stores Rev. <u>55</u> (4): 8, 10 (1945). AT-FA Jour. <u>7</u> (9): 12-3 (1945).   458

Working trees for naval stores. A. R. Shirley. Georgia Univ. Georgia Agr. Ext. Serv. Bul. 532. 46 pp. (1946).   473

x Diagrammatic sketch of continuous still for pine gum and auxiliary equipment, Naval Stores Station, Olustee, Florida (drawing). U. S. Bur. of Agr. and Indus. Chem. (1946).   478

x Flash still details, flash chamber and eight inch stripping column, Naval Stores Station, Olustee, Florida. (drawing) NS-208. (1946).   482

x Continuous still, developed for producing rosin, turpentine. Mfrs. Rec. <u>116</u> (1): 74-5, 160-1, 4 (1947).   484

x Turpentine separator with dehydrator, Revised (drawing), G. P. Shingler and E. L. Patton. NS-70. (1947).   489

x Pine gum, processing, labor and fuel costs. H. P. Poole and E. L. Patton. Chem. Engin. <u>54</u> (8): 102-3 (1947).   491

x Turpentine dehydrator, Revised (drawing). G. P. Shingler and E. L. Patton. NS-52. (1947).   492

Two Research Projects on Gum Naval Stores. Press Release. October 8, 1947.   496

x Modern techniques for processing pine gum. E. L. Patton. Forest Products Res. Soc. Preprint (1947). Naval Stores Rev. <u>57</u> (35): 13, 26 (1947).   498

PRODUCTION (continued)

	Reference Number
x Naval stores. E. L. Patton. Indus. and Eng. Chem. 40 (6): 995-7 (1948).	503
x Continuous distillation of gum turpentine. E. L. Patton and G. P. Shingler. Indus. and Engin. Chem. 40 (9): 1695 (1948).	504
x New naval stores product is ready for commercial evaluation. E. L. Patton. Naval Stores Rev. 59 (48): 16 (1950).	519
x New methods improve turpentine. E. P. Waite and N. C. McConnell. Naval Stores Rev. 59 (52): 18, 22, 23 (1950).	520
x Larger turpentine cups prove more efficient without effect on product yields or grade. Ralph W. Clements and D. N. Collins. Naval Stores Rev. 60 (13): 16-18 (1950). (With United States Forest Service.)	526
Still Better (staff-industry collaborative report). Will H. Shearon, Jr., E. L. Patton, E. P. Waite, and N. C. McConnell. Indus. and Engin. Chem. 42 (11): 18a-20a (1950). (With Indus. and Engin. Chem.)	528
x Better grades of rosin by proper handling of chips. N. C. McConnell. Naval Stores Rev. 60 (44): 11 (1951); AT-F... 13 (4): (1951).	530
x Improved filtering process developed for gum naval stores processing. Hugh B. Summers, Jr. and E. P. Waite. Naval Stores Rev. 60 (53): 12-13 (1951); AT-FM. 13 (6): 16-17 (1951).	533
x Better rosin by double-washing in gum cleaning. N. C. McConnell and E. L. Patton. AIC-311. (Processed.) (July 1951).	534
x Progress in gum naval stores processing. E. L. Patton. Jr. of Southern Research Vol. III, No. 6, pp. 20-21, Nov.-Dec. 1951.	543
x Promising new pine gum process. E. P. Waite, D. N. Collins, and H. B. Summers, Jr. Chem. Engin. Vol. 52, No. 2, pp 109 and 201, February 1952.	545

	Reference Number
USES	

x Symposium on rosin, a discussion of the uses of rosin in the industries and the production and properties of gum rosin and wood rosin. A. S. T. M. Proc. 26 (II): 493-534 (1926).   81

A new non-crystallizing gum rosin. S. Palkin and W. C. Smith. Oil and Soap 15: 120-2 (1938).   276

x Gum and wood rosin - their service in great consuming industries - as physical and chemical properties are better known uses in industry should increase. U. S. Bur. of Chem. and Soils. Gamble's International Naval Stores Year Book 1938-39 pp. 126-8 (1938).   282

x Uses of turpentine and rosin. U. S. Bur. of Agr. Chem. and Engin. Mimeographed. MC-40. 5 pp. (1938).   293

x The use of copper resinate as a treatment for paper pots. M. Leatherman and V. R. Boswell. Amer. Soc. Hort. Sci. Proc. 37: 951-5 (1939).   301

x Utilization of naval stores. C. F. Speh. Indus. and Engin. Chem. 31: 166-8 (1939).   305

x Rosin...its possible use in the present emergency as a partial replacement for coconut oil. W. D. Pohle. U. S. Bur. of Agr. Chem. and Engin. Mimeographed. ACE-162. 2 pp. (1942). Soap Sanit. Chem. 18 (2): 29, 69 (1942).   390

COMPOSITION, SPECIFICATIONS, AND ANALYTICAL METHODS	Reference Number
The detection and estimation of coal-tar oils in turpentine. V. E. Grotlisch and W. C. Smith. Indus. and Engin. Chem. 13 (9): 791-3 (1921).	42
Changes in powdered rosin stored in closed containers. F. P. Veitch and W. F. Sterling. Indus. and Engin. Chem. 15 (6): 576 (1923).	55
Tentative method of test for determination of toluol insoluble matter in rosin (chiefly sand, chips, dirt and bark). D269-27T. A. S. T. M. Proc. 27 (I): 895 (1927). (See A. S. T. M. Standards 44 (III): 1003 (1944)).	86
Proposed method for determination of softening and fluid temperatures of rosin (capillary tube method). A. S. T. M. Proc. 28 (I): 608-9 (1928). (For ball and ring method of determining softening point of rosin, see A. S. T. . Standards 44 (III): 2119 (1944)).	98
Proposed method of test for softening temperature of rosin (ring-and-ball method). A. S. T. M. Proc. 28 (I): 610-12 (1928).	99
Determination of small quantities of sulfur and chlorine when present in turpentine. W. C. Smith. Indus. and Engin. Chem., Analyt. Ed. 3: 354-5 (1931).	135
x The fractionation of American gum spirits of turpentine and evaluation of its pinene content by optical means. S. Palkin. U. S. Dept. of Agr. Tech. Bul. 276. 13 pp. (1932).	139
Specific gravity of gum spirits turpentine. W. C. Smith and G. P. Shingler. Naval Stores Rev. 42 (12): 14 (1932).	145
Standard methods of sampling and testing turpentine. D233-33. A. S. T. M. Standards 33 (II): 668-75 (1933). (See A. S. T. . Standards 44 (II): 1501 (1944)).	155
Standard specifications for spirits of turpentine. D13-34. A. S. T. M. Standards 34: 111-2 (1934). (See A. S. T. M. Standards 44 (II): 963 (1944)).	172
Acidity titration of low-grade rosins. W. C. Smith. Indus. and Engin. Chem., Analyt. Ed. 6: 122-3 (1934).	176
Analytical methods for rosin. F. P. Veitch. Assoc. Off. Agr. Chem. Jour. 18: 66-9 (1935).	192

Report on naval stores. F. P. Veitch. Assoc. Off. Agr. Chem. Jour. <u>19</u>: 390-2 (1936).   231

x The composition of American steam-distilled wood turpentine and a method for its identification. T. C. Chadwick and S. Palkin. A. S. T. M. Proc. <u>37</u> (II): 574-81 (1937).   240

x Influence of solvent on the saponification number of rosin. W. C. Smith. Indus. and Engin. Chem., Analyt. Ed. <u>9</u>: 469-71 (1937).   263

x Composition and fractionation of American steam distilled wood turpentine. S. Palkin, T. C. Chadwick, and . B. Matlack. U. S. Dept. of Agr. Tech. Bul. 596. 29 pp. (1937).   268

x Titre of rosin-fatty acid mixtures. W. D. Pohle. Soap <u>16</u> (3): 61 (1940).   330

x Hydrometer for turpentine indicating pounds per gallon. W. C. Smith. Indus. and Engin. Chem., Analyt. Ed. <u>13</u>: 112-4 (1941).   359

x Composition of American gum turpentine exclusive of the pinenes. T. C. Chadwick and S. Palkin. U. S. Dept. of Agr. Tech. Bul. 749. 16 pp. (1941).   362

x Determination of unsaturation in the terpene series. L. M. Joshel, S. A. Hall and S. Palkin. Indus. and Engin. Chem., Analyt. Ed. <u>13</u>: 447-9 (1941).   371

x Content of l-pimaric acid in pine oleoresin, improved methods for its determination. E. E. Fleck and S. Palkin. Indus. and Engin. Chem., Analyt. Ed. <u>14</u>: 146-7 (1942).   393

COMPOSITION, SPECIFICATIONS, AND ANALYTICAL METHODS (continued)

Reference Number

x l-Pimaric acid content of longleaf and slash pine oleoresins. B. L. Davis and E. E. Fleck. Indus. and Engin. Chem. <u>35</u>: 171-2 (1943).   420

SCIENTIFIC AND TECHNICAL

	Reference Number
Improved gauze-plate laboratory rectifying column. S. Palkin. Indus. and Engin. Chem., Analyt. Ed. $\underline{3}$: 377-8 (1931).	137
The resin acids of American turpentine gum. The preparation of the pimaric acids from pinus palustris. S. Palkin and T. H. Harris. Amer. Chem. Soc. Jour. $\underline{55}$: 3677-84 (1933).	168
x Oleoresin from individual trees of slash and longleaf pine. A. P. Black and S. M. Thronson. Indus. and Engin. Chem. $\underline{26}$: 66-9 (1934).	173
A differential pressure control mechanism for vacuum distillation. S. Palkin and O. A. Nelson. Indus. and Engin. Chem., Analyt. Ed. $\underline{6}$: 386-7 (1934).	186
x The preparation of l-abietic acid (Schulz) and properties of some of its salts. S. Palkin and T. H. Harris. Amer. Chem. Soc. Jour. $\underline{56}$: 1935-7 (1934).	188
x A precision oil gage. S. Palkin. Indus. and Engin. Chem., Analyt. Ed. $\underline{7}$: 434-5 (1935).	207
x Improvements in design of pressure control assembly. S. Palkin. Indus. and Engin. Chem., Analyt. Ed. $\underline{7}$: 436 (1935).	209
A catalytic method for the preparation of α-pyroabietic acid. E. E. Fleck and S. Palkin. Science $\underline{85}$ (2196): 126 (1937).	243
Catalytic isomerization of the acids of pine oleoresin and rosin. E. E. Fleck and S. Palkin. Amer. Chem. Soc. Jour. $\underline{59}$: 1593-5 (1937).	259
On the nature of pyroabietic acids. E. E. Fleck and S. Palkin. Amer. Chem. Soc. Jour. $\underline{60}$: 921-5 (1938).	275
A simplified precision oil manometer. T. C. Chadwick and S. Palkin. Indus. and Engin. Chem., Analyt. Ed. $\underline{10}$: 399-400 (1938).	284
x The dihydroabietic acids from so-called pyroabietic acids. E. E. Fleck and S. Palkin. Amer. Chem. Soc. Jour. $\underline{60}$: 2621-2 (1938).	289

SCIENTIFIC AND TECHNICAL (continued)

		Reference Number
x	The composition of so-called pyroabietic acid prepared without catalyst. E. E. Fleck and S. Palkin. Amer. Chem. Soc. Jour. <u>61</u>: 247-9 (1939).	304
x	The presence of dihydroabietic acid in pine oleoresin and rosin. E. E. Fleck and S. Palkin. Amer. Chem. Soc. Jour. <u>61</u>: 1230-2 (1939).	307
	A device to prevent bumping and promote boiling. S. Palkin and T. C. Chadwick. Indus. and Engin. Chem., Analyt. Ed. <u>11</u>: 509-10 (1939).	316
x	Lactonization of dihydro-l-abietic and dihydro-l-pimaric acids. E. E. Fleck and S. Palkin. Amer. Chem. Soc. Jour. <u>61</u>: 3197-9 (1939).	319
x	A study of factors influencing the color contributed to soap by gum rosin. W. D. Pohle and C. F. Speh. Oil and Soap <u>17</u>: 100-6 (1940).	333
x	Surface tension of rosin soap solutions. W. D. Pohle. Oil and Soap <u>17</u>: 150-1 (1940).	341
x	A hydroxy-lactone from d-pimaric acid. E. E. Fleck and S. Palkin. Amer. Chem. Soc. Jour. <u>62</u>: 2044-7 (1940).	342
x	Detergent action of rosin soaps and fatty acid -- rosin soaps. W. D. Pohle and C. F. Speh. Oil and Soap <u>17</u>: 214-6 (1940).	344
x	The germicidal action of cleaning agents - a study of a modification of Price's procedure. W. D. Pohle and L. S. Stuart. Jour. Infect. Dis. <u>67</u>: 275-81 (1940).	350
x	The germicidal activity of rosin soap and fatty acid - rosin soap as indicated by hand-washing experiments. W. D. Pohle and L. S. Stuart. Oil and Soap <u>18</u>: 2-7 (1941).	356
x	Germicidal activity of soaps. L. S. Stuart and W. D. Pohle. Soap and Sanit. Chem. <u>17</u> (2): 34-7, 73-4; <u>17</u> (3): 34-7, 73-4 (1941).	363
	Heat transfer coefficients for the condensation of mixed vapors of turpentine and water on a single horizontal tube. E. L. Patton and R. A. Feagan, Jr. Indus. and Engin. Chem. <u>33</u>: 1237-9 (1941).	375

SCIENTIFIC AND TECHNICAL (continued)

	Reference Number
A method of installing tube wall thermocouples. E. L. Patton and R. A. Feagan, Jr. Indus. and Engin. Chem., Analyt. Ed. 13: 823-4 (1941).	376
Heat requirements for steam distillation of turpentine gum. E. L. Patton and R. A. Feagan, Jr. Indus. and Engin. Chem. 33: 1380-1 (1941).	377
x Vapor phase thermal isomerization of α- and β-pinene. L. A. Goldblatt and S. Palkin. Amer. Chem. Soc. Jour. 63: 3517-22 (1941).	380
x Foaming properties of rosin soap and their comparison with those of fatty acid soaps. W. D. Pohle. Oil and Soap 18: 247-8 (1941).	382
x Solubility of calcium soaps of gum rosin, rosin acids and fatty acids. W. D. Pohle. Oil and Soap 18: 244-5 (1941).	383
x The oxidation of β-pinene with selenium dioxide. L. Joshel and S. Palkin. Amer. Chem. Soc. Jour. 64: 1008 (1942).	398
x Ester gums from rosin and modified rosins. W. D. Pohle and W. C. Smith. Indus. and Engin. Chem. 34: 849-52 (1942).	406
An oil manometer-manostat to control column throughput. S. A. Hall and S. Palkin. Indus. and Engin. Chem., Analyt. Ed. 14: 652-4 (1942).	409
x An efficient column suitable for vacuum fractionation, concentric tube type. S. A. Hall and S. Palkin. Indus. and Engin. Chem., Analyt. Ed. 14: 807-11 (1942).	412
x A tilting arc flow divider suitable for reflux ratio control. S. Palkin and S. A. Hall. Indus. and Engin. Chem., Analyt. Ed. 14: 901-2 (1942).	413
A light-and-shadow box as a visual aid in measuring spectra. J. J. Hopfield. Optical Soc. Amer. Jour. 33: 113-4 (1943).	421
Raman spectra of two forms of allo-ocimene. J. J. Hopfield, S. A. Hall, and L. A. Goldblatt. Amer. Chem. Soc. Jour. 66: 115-8 (1944).	435

SCIENTIFIC AND TECHNICAL (continued)

Reference Number

x The production of α- and β-pyronene from allo-ocimene. L. A. Goldblatt and S. Palkin. Amer. Chem. Soc. Jour. 66: 655-6 (1944). 439

x The continuous thermal isomerization of α-pinene in the liquid phase. T. R. Savich and L. A. Goldblatt. Amer. Chem. Soc. Jour. 67: 2027-31 (1945). 463

x Production of isoprene from turpentine derivatives. B. L. Davis, L. A. Goldblatt, and S. Palkin. Indus. and Engin. Chem. 38: 53-7 (1946). 465

x Di- and triethylene glycols as manostat fluids. W. J. Runckel and D. M. Oldroyd. Indus. and Engin. Chem., Analyt. Ed. 18: 80-1 (1946). 466

x Viscosity of pine gum. W. J. Runckel and I. E. Knapp. Indus. and Engin. Chem. 38: 555-6 (1946). 472

x Inhibition of myrcene polymerization during storage. W. J. Runckel and L. A. Goldblatt. Indus. and Engin. Chem. 38: 749-51 (1946). 474

x Synthesis of the monoacid chloride and the monoalkyl esters of the maleic acid anhydride addition product of 1-pimaric acid. M. M. Graff. Amer. Chem. Soc. Jour. 68: 1937-8 (1946). 476

x Effect of uniform versus intermittent product withdrawal from distillation columns. D. M. Oldroyd and L. A. Goldblatt. Indus. and Engin. Chem., Analyt. Ed. 18: 761-3 (1946). 480

x Properties of oleoresins, rosins and turpentines from chemically stimulated slash and longleaf pines. L. W. Mans and M. C. Schopmeyer. Indus. and Engin. Chem. 39 (11): 1504-6 (1947). 499

x Emulsion polymerization of myrcene. A. J. Johanson, F. L. McKennon, and L. A. Goldblatt. Indus. and Engin. Chem. 40 (3): 500-2 (1948). 501

x Emulsion copolymerization of isoprene and styrene. A. J. Johanson and L. A. Goldblatt. Indus. and Engin. Chem. 40 (11): 2086 (1948). 506

SCIENTIFIC AND TECHNICAL (continued)

Reference Number

x The properties of longleaf pine oleoresin as affected by tree characteristics and management practices. C. K. Clark and J. G. Osborne. Mimeographed. ATC-191. (12): 17 pp. (1948). (A Joint Publication by the Bureau of Agricultural and Industrial Chemistry and Southeastern Forest Experiment Station, Forest Service).   507

x Emulsifiers for GR-S from resin acid derivatives. F. L. McKennon, A. J. Johanson, E. T. Field, and R. V. Lawrence. Indus. and Engin. Chem. $\underline{41}$ (6): 1296-8 (1949).   513

x Method for identifying isobutylene. Edwin D. Parker and L. A. Goldblatt. Analyt. Chem. $\underline{21}$: 807 (1949).   514

x Synthetic rubber produced from turpentine. Res. Achvt. Sheet 124 (C). (Processed.) (December 1949).   517

x The thermal isomerization of allo-ocimene. Edwin D. Parker and L. A. Goldblatt. Amer. Chem. Soc. Jour. $\underline{72}$ (5): 2151-59 (1950).   522

x The peroxide-catalyzed addition of carbon tetrachloride to beta-pinene. Dorothy M. Oldroyd, G. S. Fisher, and L. A. Goldblatt. Amer. Chem. Soc. Jour. $\underline{72}$ (6): 2407-10 (1950).   524

x Peroxides from turpentine as catalysts for 5° C. GR-S polymerization. G. S. Fisher, L. A. Goldblatt, I. Kniel, and A. D. Snyder. Indus. and Engin. Chem. $\underline{43}$ (3): 571-74 (1951). (With Government Laboratories, University of Akron).   532

x USDA Research develops new chemicals from turpentine. Press release February 8, 1952.   544

x Fused zinc resinates from aldehyde-modified rosin. W. E. St. Clair and R. V. Lawrence. Indus. and Engin. Chem. Vol. 44, No. 2, pp. 349-351, February 1952.   547

x Navy and Agriculture Chemists make synthetic lubricants from turpentine. Dept. of Defense, Office of Public Inf. Press release. April 16, 1952.   548

American Turpentines. American Oil of Turpentine--Gum Spirits of Turpentine. L. A. Goldblatt. Guenther's Essential Oils, Vol. VI (1952) pp. 253-308.   550

GENERAL AND MISCELLANEOUS

	Reference Number
The Federal Naval Stores Act. F. P. Veitch. Indus. and Engin. Chem. **16**: 640-1 (1924).	67
x What's ahead in naval stores. H. G. Knight. Mfrs. Rec. **105** (8): 34-5, 62 (1936).	230
Naval store's greatest need - scientific research. C. F. Speh. Fla. Chemurg. Conf. Proc. **57**: 49-54 (1937).	246
x The Naval Stores Research Division, its objectives, program and progress. F. P. Veitch and C. F. Speh. U. S. Bur. of Chem. and Soils. Mimeographed. C-21. 28 pp. (1937).	267
Some interesting facts about the marketing of gum naval stores. G. P. Shingler and J. O. Boynton. Naval Stores Rev. **48** (7): 4, 15 (1938).	277
x Naval stores studies. C. F. Speh. Mfrs. Rec. **109** (2): 30-1, 54 (1940).	327
The Naval Stores Research Division and the Naval Stores Industry. C. F. Speh. Gamble's International Naval Stores Yearbook **1940-41**, pp. 129-32 (1940).	339
The work of the Naval Stores Cooperative Agents of Georgia and Florida. C. F. Speh. Naval Stores Rev. **50** (21): 8, 13 (1940).	343
x Some publications relating to production, properties, examination, marketing and uses of naval stores (Revised). U. S. Bur. of Agr. Chem. and Engin. Mimeographed. ACE-74. 5 pp. (1941).	357
Turpentine in the war effort. S. Palkin. Mfrs. Rec. **112** (5): 30-1, 62, 64, 66 (1943).	425
Contributions of Naval Stores Research Division to the processing of pine oleoresin. G. P. Shingler. AT-FA Jour. **8** (6): 8-9, 20 (1946).	469
x Selected publications of the Naval Stores Research Division on production, properties, and uses of naval stores. U. S. Bur. Agr. and Indus. Chem. Mimeographed. AIC-144.	483
x Hints on laboratory technique. M. . Graff. Jour. Chem. Ed. **24**: 182 (1947).	486

GENERAL AND MISCELLANEOUS (continued)   Number

Recent achievements in naval stores research by the Naval Stores Research Division. G. P. Shingler. Naval Stores Rev. International Yearbook, 1948. pp. 100-3 (1948).   505

x Applied research on pine gum at the Naval Stores Station. D. N. Collins. AIC-251. (Processed.) (August 26, 1949).   515

Role of Olustee gum cleaning in a new pine gum industry. E. L. Patton and G. P. Shingler. Naval Stores Rev. Internatl. Yearbook. 106-109 (1949).   516

x Latest methods applied to naval stores processing. F. L. McKennon and E. P. Waite. Paint, Oil, and Chem. Rev. 113 (12): 17, 18, 20 (1950).   525

x Research in the naval stores industry. E. L. Patton and F. L. McKennon. Naval Stores Rev., Internatl. Yearbook. 78-80 (1950).   527

x Research modernizes the gum naval stores industry. E. L. Patton. Naval Stores Rev. 60 (48): 13, 14, 22, 23 (1951).   531

Past, present and future of a rejuvenated process industry. A new tack for gum naval stores. D. N. Collins and E. L. Patton. Chem. Engin. 58 (9): 154-156 (1951).   537

x New horizons seen for pine gum through research. E. L. Patton. Naval Stores Rev., Internatl. Yearbook. 80-82, 96, 97 (1951).   539

x The industrial utilization of rosin. R. V. Lawrence. Yearbook of Agric. 1950-1951. Crops in Peace and War. 822-826; Naval Stores Rev. 61 (30): 16, 17, 25-28 (1951).   540

x The chemicals we get from turpentine. L. A. Goldblatt. Yearbook of Agric. 1950-1951. Crops in Peace and War. 814-821.   541

x RESEARCH-Naval Stores Bonus. Chem. Week. 68 (19): 33 (1951). (Editorial-Malco-pimaric acid-a new product.)   535

x Research-The key to progress in naval stores utilization. E. L. Patton. FLACS-pp. 3, 5, 6, 8, 9, 11. February 1952.   546

x Physical and Chemical Properties of Rosin. Mimeographed.   549

	Reference Number
PATENTS	

ethod for making varnish materials. W. F. Sterling, V. E. Grotlisch, and F. P. Veitch. U. S. Patent No. 1,395,874. November 1, 1921. 43

Steam turpentine still. J. O. Reed. U. S. Patent No. 1,667,168. April 24, 1928. 104

Turpentine gum sampler. N. C. McConnell. U. S. Patent No. 1,953,886. April 3, 1934. 178

Refining of natural oleoresin. R. W. Frey and W. C. Smith. U. S. Patent No. 2,039,481. May 5, 1936. 226

Gauge for measurement of gas pressures. S. Palkin. U. S. Patent No. 2,051,740. August 18, 1936. 232

Turpentine dip barrel head lock and remover. L. Evans. U. S. Patent No. 2,052,223. August 25, 1936. 233

Process for separating pine gum into fractions. S. Palkin and T. H. Harris, Jr. U. S. Patent No. 2,086,777. July 13, 1937. 256

Gauge for measurement of gas pressures. S. Palkin and T. C. Chadwick. U. S. Patent No. 2,169,812. August 15, 1939. 313

Noncrystallizing gum rosin. S. Palkin and W. C. Smith. U. S. Patent No. 2,176,660. October 17, 1939. 318

Multiple filter device. S. Palkin. U. S. Patent No. 2,198,175. April 23, 1940. 332

Stable rosin acid, rosin ester, and rosin product containing them and a method for their production. E. E. Fleck and S. Palkin. U. S. Patent No. 2,239,555. April 22, 1941. 367

Process for gum refining. W. C. Smith, J. O. Reed, F. P. Veitch, and G. P. Shingler. U. S. Patent No. 2,254,785. September 2, 1941. 374

Filtering apparatus. J. O. Reed. U. S. Patent No. 2,272,583. February 10, 1942. 392

Thixotropic composition of matter. M. Leatherman. U. S. Patent No. 2,277,048. March 24, 1942. 395

Process for stabilizing rosin and pine oleoresin. W. D. Pohle and W. C. Smith. U. S. Patent No. 2,277,351. March 24, 1942. 396

PATENTS (continued)

	Reference Number
Filter apparatus. S. Palkin. U. S. Patent No. 2,285,048. June 2, 1942.	403
Process for producing rosin. J. O. Reed. U. S. Patent No. 2,295,235. September 8, 1942.	410
Process for refining oleoresin. J. O. Reed. U. S. Patent No. 2,307,078. January 5, 1943.	417
Process for refining turpentine. J. O. Reed. U. S. Patent No. 2,308,715. January 19, 1943.	419
Hanger. J. O. Reed. U. S. Patent No. 2,316,103. April 6, 1943.	424
Fractionating apparatus. J. O. Reed. U. S. Patent No. 2,319,365. May 18, 1943.	426
Apparatus for washing oleoresin. J. O. Reed. U. S. Patent No. 2,322,252. June 22, 1943.	428
Fireproofing composition. M. Leatherman. U. S. Patent No. 2,326,233. August 10, 1943.	429
Fractionating column and scrubbing tower. S. Palkin and S. A. Hall. U. S. Patent No. 2,344,360. March 21, 1944.	438
Apparatus for melting and processing crude oleoresin. J. O. Reed. U. S. Patent No. 2,356,798. August 29, 1944.	445
Process for refining pine oleoresin. E. E. Fleck. U. S. Patent No. 2,359,980. October 10, 1944.	447
Cooker and dehydrator. J. O. Reed. U. S. Patent No. 2,361,151. October 24, 1944.	448
Production of rosin and turpentine. J. O. Reed. U. S. Patent No. 2,363,692. November 28, 1944.	450
Process for treating crude pine tars. W. C. Smith. U. S. Patent No. 2,364,104. December 5, 1944.	451
Apparatus for precision control of flow division for fractionating columns and the like. S. Palkin and S. A. Hall. U. S. Patent No. 2,369,913. February 20, 1945.	455

PATENTS (continued)

Reference Number

Dehydrator. J. O. Reed. U. S. Patent No. 2,370,422. February 27, 1945. — 456

Method for refining and distilling oleoresin. J. O. Reed. U. S. Patent No. 2,377,183. May 29, 1945. — 457

Process for treating crude pine tar. W. C. Smith. U. S. Patent No. 2,379,662. July 3, 1945. — 460

Process for refining crude oleoresin. R. V. Lawrence. U. S. Patent No. 2,395,190, February 19, 1946. — 468

Process for refining oleoresin. R. V. Lawrence. U. S. Patent No. 2,411,925. December 3, 1946. — 479

Process for converting nopinene to myrcene. L. A. Goldblatt and S. Palkin. U. S. Patent No. 2,420,131. May 6, 1947. — 487

Process for the isomerization of allo-ocimene. L. A. Goldblatt. U. S. Patent No. 2,427,497. September 16, 1947. — 494

Process for depolymerizing allo-ocimene polymers. D. M. Oldroyd, T. R. Savich and L. A. Goldblatt. U. S. Patent No. 2,427,506. September 16, 1947. — 495

Drying rosin oil and method of producing same. E. E. Fleck. U. S. Patent No. 2,429,264. October 21, 1947. — 497

Bark chipping hack for turpentining trees. Albert G. Snow, Jr. and Hubert R. Lanier. U. S. Patent No. 2,434,869. January 20, 1948. — 500

Process for pyrolyzing alpha-pinene to allo-ocimene in liquid phase. Theodore R. Savich and Leo A. Goldblatt. U. S. Patent No. 2,437,759. March 16, 1948. — 502

Polymerizing the glycerol ester of levopimaric acid-maleic anhydride adduct. Elmer E. Fleck. U. S. Patent No. 2,458,772. January 11, 1949. — 508

Butadiene emulsion polymerization in the presence of levopimaric acid-maleic anhydride addition product esters. Ray V. Lawrence. U. S. Patent No. 2,465,888. March 29, 1949. — 509

PATENTS (continued)

Reference Number

Butadiene emulsion polymerization in the presence of levopimaric acid-maleic anhydride addition product. F. L. McKennon and Ray V. Lawrence. U. S. Patent No. 2,465,901. March 29, 1949. — 510

The production of mono-alkyl esters of the addition product of levopimaric acid with maleic anhydride. Morris M. Graff. U. S. Patent No. 2,467,126. April 12, 1949. — 511

The flash distillation of turpentine. N. C. McConnell, L. W. Mims, Harry P. Poole, and Hubert R. Lanier. U. S. Patent No. 2,500,194. March 14, 1950. — 518

Modified rosin esters. R. V. Lawrence and Muriel W. Kaufman. U. S. Patent No. 2,504,989. April 25, 1950. — 521

Process for producing myrcene from beta-pinene, Theodore Savich and L. A. Goldblatt. U. S. Patent No. 2,507,546. May 16, 1950. — 523

Organochlorosilanes processes for their production. L. A. Goldblatt and Dorothy M. Oldroyd. U. S. Patent No. 2,533,240. December 12, 1950. — 529

Halogenated terpene addition compounds. L. A. Goldblatt and Dorothy M. Oldroyd. U. S. Patent No. 2,564,685. August 21, 1951. — 536

Aluminum resinates and methods of preparation. W. E. St. Clair and R. V. Lawrence. U. S. Patent No. 2,567,250. September 11, 1951. — 538

New and improved metal resinates and method of preparation. W. E. St. Clair and R. V. Lawrence. U. S. Patent No. 2,572,071. October 23, 1951. — 542

STATISTICS

x   Annual reports on the production, distribution, consumption, and stocks of naval stores. Mimeographed cir. Beginning in 1935, these reports have been issued in May of each year, covering the preceding naval stores crop year, i. e. from April 1 through March 31.

x   Semi-annual reports on the production, distribution, consumption, and stocks of naval stores. Mimeographed.cir. Beginning in 1936, these reports have been issued in November of each year, covering the first six months of the naval stores crop year, i. e. from April 1 through September 30.

x   Quarterly reports on the production, distribution, consumption, and stocks of naval stores. Mimeographed cir. Beginning in 1942, these reports have been issued twice each year, one in August covering the preceding quarter (April 1 through June 30), and one in February covering the preceding quarter (October 1 through December 31).

NOTE

The last statistical report on naval stores issued by the Naval Stores Research Division covers the period April 1, 1946 through March 31, 1947. As of April 1, 1947, the responsibility for compiling and publishing naval stores statistics on production, consumption, and stocks was transferred to the Bureau of Agricultural Economics, Washington, D. C. Requests for reports subsequent to April 1, 1947, should be directed to the Bureau of Agricultural Economics.

Selected publications of the Naval Stores Research Division on pro
United States Bureau of Agricultural and Industrial Chemistry
CAT31114090
U S Department of Agriculture, National Agricultural Library
[26] selectedpublicat144unit_0
Jan 13, 2014

CPSIA information can be obtained
at www.ICGtesting.com
Printed in the USA
LVHW080854301118
598763LV00010B/152/P